The Pharmaceutical and GMP Dictionary

www.validationresources.org

ISBN-13: 978-1545133309
ISBN-10: 1545133301

A

Accelerated Aging

When the deterioration of a device or product component from natural aging is accelerated and simulated in the laboratory.

Accuracy

Accuracy or trueness. An expression of the closeness of agreement between the value that is accepted, either as a conventional true value or an accepted reference value and the value obtained. A system with low bias implies good accuracy and vice versa.

Adverse Event

A situation or condition that occurs when a data point, result, or process etc. is outside the expected or predetermined limits or ranges.

Affinity Chromatography

A chromatography separation method based on a chemical interaction specific to the target species. Types of affinity methods are: biosorption -site recognition (e.g., monoclonal antibody, protein A); hydrophobic interaction -contacts between non-polar regions in aqueous solutions.

Air Exchange Rate Per Hour (ACPH)

The rate of air exchange expressed as number of air changes per hour and calculated by dividing the volume of air delivered in the unit of time by the volume of space.

Active Pharmaceutical Ingredient

Any substance or mixture of substances intended to be used in manufacturing a drug (medicinal) product and that, when used in the production of a drug, becomes an active ingredient of the drug product. Such substances are intended to furnish pharmacological activity or other direct effect in the diagnosis, cure, mitigation, treatment, or prevention of disease, or to affect the structure and function of the body. (ICH Q7A, Annex 18, Part II)

Amino acid

The building block of proteins. The messenger RNA tells the cell what amino acids are needed and what order they must be arranged in to build a particular protein. There are 20 different amino acids used in the human body.

Amniocentesis

A procedure used in prenatal diagnosis to look at the chromosomes of the developing fetus. A flexible needle is inserted into the mother's uterus through the abdomen to remove a sample of the fluid surrounding the fetus (amniotic fluid). This sample can then be analysed by karyotype to look for changes in the chromosomes. The procedure can be done after 15 weeks of pregnancy. There is a 0.5% risk of miscarriage associated with this procedure, which means one in 200 women will miscarry following this procedure.

ANOVA (Analysis of Variance)

A statistical method used to evaluate the significance of differences in means due to different factor-level combinations.

ANSI

American National Standards Institute

Antimicrobial Resistance

Antimicrobial resistance corresponds to the emergence and spread of microbes that are resistant to cheap and effective first-choice, or "first-line" antimicrobial drugs.

Antigen

Substance, usually a foreign protein or carbohydrate, which when introduced into a organism, activates specific receptors on the surface immunocompetent T and B lymphocytes. After interaction between antigen and receptors, there usually will be induction of an immune response, i.e. production of antibodies capable of reacting specifically with determinant sites on the antigen.

Application

A term most often used in relation to Software validation and computerized systems. It is any software installed on a defined platform providing specific functionality.

Approve

"Approve" the device after reviewing a premarket approval (PMA) application that has been submitted to FDA.

Alert limit

A value reached when the normal operating range of a critical parameter has been exceeded, indicating that corrective measures may need to be taken to prevent the action limit being reached.

Assay

A method for determining the presence or quantity of a component.

At-rest

A room (HVAC) condition where the installation is complete with equipment installed and operating in a manner agreed upon by the customer and supplier, but with no personnel present.

AVL (Approved vendor list)

A list of all the vendors or suppliers approved by a company as sources from which to purchase materials.

Artwork

Electronic files or printouts containing the representation of a packaging item, graphical elements, and regulatory text. Approved artworks are used by suppliers for printing.

Aseptic (conditions)

Conditions in the working environment under which the potential for microbial and/or viral contamination is minimized.

ASTM

American society for testing and materials.

ATEX

ATEX, an acronym of the French Atmospheriques Explosives. This European Directive amends and adds safety requirements for hazardous areas in the relevant national legislation in the member states of the European Union, bringing in a common standard. Where equipment is to be used in potentially explosive atmospheres containing gas or combustible dust, it must comply with the ATEX directive.

Audit Trail

The audit trail is a control mechanism of a system that allows all data entered or modified to be traced back to the original data. A reliable and secure audit trail is particularly important in conjunction with the creation, change or deletion of GMP relevant electronic records.

Automated System

Term used to cover a broad range of systems, including automated manufacturing equipment, control systems, automated laboratory systems, manufacturing execution systems and computers running laboratory or manufacturing database systems. The automated system consists of the hardware, software and network components, together with the controlled functions and associated documentation. Automated systems are sometimes referred to as computerised systems; in this guide the two terms are used interchangeably.

Adventitious Organism

Bacteria, yeast, mold, mycoplasma or viruses that can potentially contaminate prokaryote or eukaryote cells used in production. Potential sources of adventitious organisms include the serum used in cell culture media, persistently or latently infected cells, or the environment.

Acceptable Quality Level (AQL)

The AQL of a sampling plan is the Process Performance Level routinely accepted by the sampling plan.

B

Bacteriophage

A virus that attacks bacteria. The lambda bacteriophage is frequently used as a vector in recombinant gene experiments.

Barrier System (Isolator)

A physical separation between a critical processing zone and the personnel operating within the other processing zones.

Basis of Design

A design document that demonstrates a thorough understanding of the project and its intended output. Typically contains preliminary drawings and system descriptions etc. Together with the URS and the Detailed Design, it provides overall evidence that the design addresses the requirements of the equipment, system or facility.

Bias

The difference between observed "average of measurements" and a reference value; also referred to as accuracy.

Bioactivity

The level of specific activity or potency as determined by animal, cell culture, or in vitro biochemical assay.

Biocompatibility

A measure of how a biomaterial interacts in the body with the surrounding cells, tissues and other factors.

Bioburden

The level and type of micro-organisms that can be present in raw materials, API starting materials, intermediates or APIs. Bioburden should not be considered contamination unless the levels have been exceeded or defined objectionable organisms have been detected.

Biofilm

A community of microbes embedded in an organic polymer matrix, adhering to a surface. (Carpentier, B.,1993)

Biological Indicators

Test system containing viable microorganisms providing a defined resistance to a specified sterilization process, e.g. Vaporised hydrogen peroxide.

Biomaterial

Any matter, surface, or construct that interacts with biological systems. Biomaterials can be derived from nature or synthetic (manufactured). The active substance of a biosimilar medicine is comparable to a biological reference medicine. Biosimilar and biological reference medicines are used at the same dose to treat the same disease. The name, appearance and packaging of a biosimilar medicine differs to that of a biological reference medicine.

Bioreactor

A vessel in which the central reactions of a biotechnological process takes place. Typically the vessel contains microbes grown under controlled conditions of temperature, aeration, mixing, acidity and sterility.

Blending

The process of combining materials within the same specification to produce a homogeneous intermediate or API. In process mixing of fractions from single batches (e.g., collecting several centrifuge loads from a single crystallisation batch) or combining fractions from several batches for further processing is considered to be part of the production process and is not considered to be blending [ICH Q7]

Bracketing

A Bracketing (aka family or matrix) approach can be used where similar products are produced using the same equipment and processes. A particular product size or product configuration may be selected to represent the worst-case product. Therefore, by qualifying the worst case, all of the other products within the family are considered validated.

Body orifice

Any natural opening in the body, as well as the external surface of the eyeball, or any permanent artificial opening, such as a stoma or permanent tracheotomy.

Borderline Classifications

In certain circumstances, it may not be clear if a product falls under the medical device legislation or whether to classify a device as a medicine, cosmetic, biocide and so on. The decision will largely depend on the particular intended use of the product, as assigned by the manufacturer, and on the demonstrated mode of action. The manufacturer's claims must be substantiated by relevant data.

Buffer

A solution that resists change in pH when an acid or alkali is added, or when solutions are diluted

Bulk Product

Any pharmaceutical form (liquid, powder, suspension) that is to be filled into either another container or its final container at the next process step; or is already filled into its final container to be labelled and packaged at the next process step.

BOM

Bill of Materials.

BSI

British Standards Institute.

C

CAD, Computer Aided Drawing

A system used to create physical designs, usually three-dimensional. Some examples of CAD software are SolidWorks, Pro/ENGINEER and AutoCAD.

Calibration

The a requirement that demonstrates a particular instrument or device produces results within specified limits by comparison with those produced by a reference or traceable standard over an appropriate range of measurements.

Campaign (Process)

A production strategy where consecutive batches of an API, a finished product, or intermediates are processed before the production line/system is cleaned.

Capability (Process Capability)

Process Capability is a measure of how capable the process is of producing product meeting specified requirements. It is a measure of the actual variation in that product characteristic compared to the product specifications. Indices are used to represent the Process Capability such as Pp, Cp and Ppk, Cpk, depending on how the data is collected e.g. multiple batches over time.

CAPA

A Corrective and preventive action. A systematic approach that includes actions needed to correct, prevent recurrence and eliminate the cause of potential nonconforming product and other quality problems (preventive action) (21CFR 820.100)

Cell Bank

A culture of cells stored in appropriate containers under defined conditions, whose contents are of uniform composition.

Cell Culture

The in- vitro growth of cells isolated from multicellular organisms. These cells are usually of one type.

Cell Differentiation

The process whereby descendants of a common parental cell achieve and maintain specialization of structure and function.

Cell Fusion

The formation of a hybrid cell with nuclei and cytoplasm from different cells, produced by fusing two cells of the same or different species.

Cell Line

Cells that acquire the ability to multiply indefinitely in-vitro.

Change Control

A formal system by which qualified representatives of appropriate disciplines review proposed or actual changes that may impact the validated status.

Change Notification (Agreement)

A signed declaration that states that the Supplier agrees to notify the customer of changes in its product or process in order to allow the customer determine whether the changes can affect the quality of finished goods or Quality System.

Change Management

An overarching approach to change control that is used manage changes to the product, process, facilities or utilities are assessed, planned and reviewed as part of a formal systematic process.

Chemical Indicator
A test device that responds to a sterilization process parameter such as VHP.

Cleaning

The process of removing potential contaminants from process equipment and maintaining the condition of equipment such that the equipment can be safely used for subsequent product manufacture.

Cleaning Agent

The chemical agent or solution used as an aid in the cleaning process.

Cleaning Validation

Documented evidence that provides a high degree of assurance that a specific cleaning process will consistently produce a result meeting predetermined requirements for cleanliness.

Cleaning Verification

Confirmation by examination and provision of objective evidence that specific requirements have been fulfilled.

Clean Hold Time (CHT)

The total time the parts or equipment are held clean post-cleaning.

Cocurrent (flow)

This is when the fluids are applied in the same direction. Cocurrent flow is less effective as less heat can be transferred, therefore it is less commonly used.

Code of Federal Regulations (CFR)

Regulations issued by U.S. government agencies. The individual titles making up the regulations are numbered the same way as the federal laws on the same topic.

Competent Authority

A competent authority is the legally designated authority mandated to monitor compliance with directives and legal requirements within the industry. The competent authority has the power to grant and revoke licenses.

Compendial Organisations

Organizations certifying material standards that meet compendial requirements and acceptance criteria. (e.g. USP).

Commissioning

An engineering activity that includes all aspects of bringing a system, piece of equipment or process is installed and ready for use. Commissioning involves both requirements of Installation Qualification (IQ) and Operational Qualification (OQ).

Computer System

A group of hardware components and associated software, designed and assembled to perform a specific function or group of functions.
[EU GMP Guide, Part II, ICH Q7]

Computerised System

A system including the input of data, electronic processing and the output of information to be used either for reporting or automatic control. [EU GMP Guide, Glossary]

Computer System Validation

A process that confirms by examination and provision of objective evidence that the computer system conforms to user needs and intended uses. System validation is a process for achieving and maintaining compliance with GxP regulations and fitness for intended use by adoption of life cycle activities, deliverables, and controls.

Concurrent Validation

Concurrent Validation occurs when activities are executed at the same time as one another or concurrent to a product launch.

Container Closure Integrity
The ability of a container closure system to provide protection and maintenance of product identity, strength, quality and purity throughout its shelf-life.

Confidence Level

Confidence Level is expressed as a percentage and represents the probability that the conclusion of the test is correct. A 95% confidence level means you can be 95% certain that the conclusion is correct.

Conflict Of Interest

A conflict of interest is a situation in which a public official's decisions are influenced by the official's personal interests.

Continual Improvement, CI

Ongoing activities to evaluate and positively change products, processes, and the quality system to increase effectiveness

Consent Decree

A consent decree is a binding order issued by a judge that stipulates the voluntary agreement by the participants in a case of litigation. Decrees are sometimes issued after one party voluntarily agrees to cease a particular action without admitting to any illegality of the action to date.

Colony Forming Unit

One or more microorganisms that produce a visible, discrete growth on an agar-based microbiological medium.

Controlled Substances

Products that are categorized due to their potential for abuse, medical use and requirement for medical supervision.

Controlled classified areas

An environment supplied with HEPA filtered air where materials, equipment, and personnel are regulated to control viable and non-viable particulates to an acceptably low level. Such areas are classified according to the maximum level of airborne particulate allowed.

CNC (Controlled Not Classified)

While these are not ISO recognized room classes, they are generally used to describe non-GMP areas with a level of control in effect.

Clear (FDA)

"clear" the device after reviewing a premarket notification, otherwise known as a 510(k) (named for a section in the Food, Drug, and Cosmetic Act), that has been filed with FDA, or

Cleanroom

An area (or room or zone) with defined environmental control of particulate and microbial contamination, constructed and used in such a way as to reduce the introduction, generation and retention of contaminants within the area.

Containment

A process or device to contain product, dust or contaminants in one zone, preventing it from escaping to another zone.

Contamination

The undesired introduction of impurities of a chemical or microbial nature, or of foreign matter, into or onto a starting material or intermediate, during production, sampling, packaging or repackaging, storage or transport.

Continued Process Verification

Once the initial validation is completed it is important that the system or process remains within the validated state. This is done by monitoring the performance and output of the system or equipment. Furthermore, any changes to this system or equipment must be assessed and documented in order to assure the product is safe and meets acceptance criteria.

Critical Aspects

Critical aspects of manufacturing systems include the functions, features, abilities, and performance or characteristics required for the manufacturing process and systems to ensure consistent product quality and patient safety. They should be identified and documented based on scientific product and process understanding.

Critical Quality Attribute, CQA (Critical-to-Quality)

A property or characteristic with specific nominal value and appropriate limit and range providing a particular quality attribute. A CQA typically is classed as a high risk requirement, where the safety or efficacy of the product depends on the CQA been within the specified limits.

CCC (Mark)

The "China Compulsory Certificate" mark, commonly known as CCC Mark, is a safety mark for many products sold on the Chinese market. As of 2013, medical devices do not require this certification.

CDC

Center for Disease Control & Prevention (USA)

CDRH

Center for Devices and Radiological Health (USA)

CE Marking

The CE Marking is a mandatory conformance mark on many products (including medical devices) placed on the single market in the European Economic Area. The CE marking certifies that a product has met EU consumer safety, health or environmental equirements. By affixing the CE marking to a product, the manufacturer declares that it meets EU safety, health and environmental requirements.

CEN

Communité Européenne des Normes (European Committee for Standardization).

Clinical Trial

Clinical Trials are conducted to allow safety and efficacy data to be collected for health interventions (e.g., drugs, diagnostics, devices, therapy protocols). These trials can take place only after satisfactory information has been gathered on the quality of the non-clinical safety, and Health Authority/Ethics Committee approval is granted in the country where the trial is taking place.
Clinical Trial Sponsor
The Clinical Trial Sponsor is responsible for the safety of subjects in a clinical trial and informs local site investigators of the true historical safety record of the drug, device or other medical treatment to be tested, and of any potential interactions of the study treatment(s) with already approved medical treatments.

Cleaning

Removal of contamination or soils from an item or surface to the extent necessary for its further processing and its intended subsequent use.

CMDCAS

Canadian Medical Devices Conformity Assessment System.

CMDR
Canadian Medical Device Regulation.

Conformity

Fulfilment of a requirement or meeting a requirement.

Conformity Assessment Body (CAB)

A body, other than a Regulatory (competent) Authority, engaged in determining whether the relevant requirements in technical regulations or standards are fulfilled.

CRO

A "Contract Research Organization", also commonly known as a "Clinical Research Organization", is a service organization that provides support to the pharmaceutical and biotechnology industries. CROs offer clients a wide range of "outsourced" pharmaceutical research services to aid in the drug and medical device research and development process.

Cytokine

Small, non- immunoglobulin proteins produced by monocytes and lymphocytes that serve as intercellular communicators after binding to specific receptors on the responding cells. Cytokines regulate a variety of biological activities.

Cytopathic Effect

Morphological alterations of cell lines produced when cells are infected with a virus. Examples of cytopathic effects include cell rounding and clumping, fusion of cell membranes, enlargement or elongation of cells, or lysis of cells.

Cytotoxic

Damaging to cells.

D

Data Authentication

Within a GxP environment, authentication refers to the approval of data (electronic signatures).E-signatures are key controls within software that prompt the user to enter a unique username and password to acknowledge a recording or action. The E-signature should create a permanent link with the electronic record that cannot be removed and can be viewed through an audit trial.

Data Integrity

Is the degree to which data is reliable and without error. Data must be accurate, attributable, contemporaneous, original, legible and available. A breach of data integrity occurs when any person manipulates or distorts data and submits the results of that data as valid.

Data handling

Any GxP task that involves creation, entry, review, approval, analysis, eporting, storage, archival, retrieval, or disposal of GxP data

Data lifecycle

A process or cycle that begins from the time of data creation to the point of use and during its retention, archival, retrieval, and eventual disposal

Data Protection
Once the data is created, the handling of the data must ensure data integrity. For electronic data, this includes access control to computer systems. Other practical restrictions can also be made such as limiting room and site access to authorised personnel.

Data Retention

The controlled storage, backup and arching of data. Retention of records may be required for several decades depending on the type of data and the regulatory requirements relating the a particular product or industry.

Dead Leg

A dead leg in the world of piping terminology refers to an area of piping where there is insufficient flow or a tendency for water build-up or stagnation.
The formal definition of a dead-leg states that
Pipelines for the transmission of purified water for manufacturing or final rinse should not have an unused portion greater in length than 6 diameters (6D rule) of the unused portion of pipe measured from the axis of the pipe in use.

Debugging

The process of locating, analysing, and correcting suspected faults or machine issues.

Design controls

Design controls are a collection of practices and procedures that are incorporated into the design and development process for a product such as a medical device. It provides a structure and clear path from user needs assessment to product delivery through a step-by-step process. Design controls ensure proper assessment of the design is completed during the design and development phase. Design controls are a requirement of quality systems such as 21 CFR Part 820 (medical devices), and for certain classes of devices and per ISO 13485 - Quality Management Systems.

Decommissioning

When a system is taken out of production service and stored in an adequate environment for potential future use.

Depyrogenation

A thermal process used to destroy or remove pyrogens (endotoxins). Typically primary packaging components such as glass vials are subject to Depyrogenation.

Detection Limit

The lowest amount of analyte in a sample that can be detected but not necessarily quantitated as an exact value for an individual analytical procedure. (Ref: ICH Q2)

Detergent

A chemical (cleaning) agent that reduces contamination

Design History File
The DHF is a repository for all of the documentation generated as a result of the design control process. The DHF serves as a complete record of the design.

Design Validation

Establishing by objective evidence that device or product specifications conform to user needs and intended use(s) defined in design documentation.

Debarment

The FDA has the authority to "disqualify," or remove, researchers from conducting clinical testing of new drugs and devices when the agency determines that the researcher has repeatedly or deliberately not followed the rules intended to protect study subjects and ensure data integrity. Further, the FDA can disqualify a clinical investigator who has repeatedly or deliberately submitted false information to the agency or study sponsor in a required report.

Under its statutory debarment authority, the agency may also ban, or "debar" from the drug industry individuals and companies convicted of certain felonies or misdemeanours related to drug products. Once individuals have been subjected to "debarment," they may no longer work for anyone with an approved or pending drug product application at FDA. Debarred companies may no longer submit abbreviated drug applications.

Design qualification (DQ)

The documented verification that the proposed design of the is suitable for the intended purpose. DQs are typical deliverables for facilities, systems and equipment and or processes.

Design Space

The multidimensional combination and interaction of input variables, e.g. material attributes, and process parameters that have been demonstrated to provide assurance of quality. Working within the design space is not considered as a change.

Dichotomous Variable

An output with only two possible values. Also known as dummy or indicator variable.

Directives

Directives are legal requirements. These must be met by manufacturers. Standard such as ISO 13485 help companies meet the requirements of directives, such as "Guidelines Relating to the Application of the Council Directive 93/42/EEC on Medical Devices."

Direct impact (system)

A system that is expected to have a direct impact on product quality. These systems are designed and commissioned in line with Good Engineering Practice (GEP) and, in addition, are subject to Qualification and Validation. Such systems include HVACs and Clean utilities such as WFI (Water-for-Injection)

Disinfectant

A chemical or physical agent that reduces bioburden

Diffusion blending

A process in which particles are reoriented in relation to one another when they are placed in random motion and interparticular friction is reduced as the result of bed expansion (usually within a rotating container). Also referred to as tumble blending.

Deviations

A deviation can be simply described as an unintended event which causes a test or verification to fail to meet expected acceptance criteria.

Degree of invasiveness

A device, which in whole or in part, penetrates inside the body either through a body orifice or through the skin surface, is invasive. Invasiveness is generally categorised as invasive of a body orifice (including the surface of the eye), surgically invasive devices and implantable devices.

Device Master Record (DMR)

A compilation of records containing the procedures and specification for a device. The contents of a DMR can contain local procedures such as SOPs and work instructions along with global or divisional specifications used to detail manufacturing processes, intermediate product or final product.

Dirty Hold Time (DHT)

The total time the parts (or equipment) are held dirty prior to cleaning.

DMAIC

Define

The first phase in the DMAIC methodology is focused on project definition. This phase is critical as it provides the basis of any problem solving activity – what is the problem or issue? What is the desired outcome or result? Many six sigma projects form a project charter as part of the definition phase. Other tools used to ensure the correct focus areas are identified from the beginning include capturing the Voice of the Customer (VOC) and creating an SIPOC table. The acronym SIPOC stands for suppliers, inputs, process, outputs, and customers.

Measure

After the define phase, the next phase is measure. It is important to measure the current state of a process, or the current issue or "state of error". Measurements will inform project teams with the facts while also creating a baseline of the process or problem. This baseline may prove critical after any changes have been implemented. Comparing the state of the problem before and after can be very helpful when illustrating data.

The measure phase cannot be skipped or delayed. As the mantra goes, if you cannot measure it, you cannot change it." Or more accurately, if you cannot measure it you will not be able to demonstrate that the changes or controls have benefited the problem or goal statement.

Analyse

During the analyse step the team should be focused on identifying the key root causes. There may be more than one cause and each one may have a varying impact on the process. Thus the analyse stage should be given adequate time to ensure all factors are considered.

Improve

With the weight of the previous steps (define, measure, analyse), the project team must distil a practical, workable and lasting improvement solution.

- Develop potential solutions
- Evaluate, select, and optimise best solutions

- Develop the next state of value stream map(s)
- Implement proof of principle of the pilot study

Control

After improvements have been identified and implanted, effective controls must be applied and sustained into the future. As with many systems in engineering, if they are not maintained they can degrade over time. So too with controls, if they are not implemented and maintained, they can simply fall away or become less effective. Above all, the purpose of any control is to ensure quality and safety of the product, and if the process is in control, the customer is satisfied.

DNA (DEOXYRIBONUCLEIC ACID)

The basic biochemical component of the chromosomes and the support of heredity. DNA contains the sugar deoxyribose and is the nucleic acid in which genetic information is stored (apart from some viruses).

DNA CLONING

Production of many identical copies of a defined DNA fragment.

DNA Polymerase

An enzyme which catalyses the synthesis of double-stranded DNA from single- stranded DNA.

DNA Synthesis

The formation of DNA by the sequential addition of nucleotide bases

DNase

An enzyme which produces single- stranded nicks in DNA. DNase is used in nick translation.

Drug Product

The dosage form in the final immediate packaging intended for marketing. The finished dosage form that contains a Drug Substance, generally, but not necessarily in association with other active or inactive ingredients. (FDA)

Duration of Contact

In determining the classification of a device the duration that the device is in continuous contact with the patient is defined as transient, short term or long term. The longer the device is in contact with the patient or user, the greater the risk and therefore this has to be taken into account when determining classification. Continuous use is defined in MEDDEV 2.4/1 as the uninterrupted actual use for the intended purpose. Where use of a device is discontinued in order that the device is immediately replaced with an identical device (e.g. replacement of a urethral catheter) this shall be considered as continuous use of the device.

D Value

The time (in minutes) or radiation dose required to reduce the microbial population by 90% or 1 log10 cycle (i.e., to surviving fraction of 1/10) and must be associated with the specific lethal conditions at which it was determined (USP <1229>)

E

E. coli (Escherichia coli)

A bacterium found in the intestinal tracts of most vertebrates. It is used extensively in recombinant DNA research because it has been genetically well characterized.

Enzymatic

Activity of an enzyme which is a substance produced by a living organism and acting as a catalyst to promote a specific biochemical reaction

Enzyme

A protein that facilitates a biochemical reaction. Many essential reactions in the body require the help of enzymes and would not proceed on their own

Enzyme-Linked Immuno Assays (EIA)

Enzyme-Linked Immuno Assays (EIA) are use to measure the amount of a particular substance by virtue of its binding to a specific antibody. Examples of EIA include ELISA and Western blotting

Enzyme-Linked Immunosorbent Assay

The ELISA is a fundamental tool of clinical immunology, and is used as an initial screen for HIV detection. Based on the principle of antibody-antibody interaction, this test allows for easy visualization of results and can be completed without the additional concern of radioactive materials use

Electronic Signatures

Electronic signatures are computer-generated character strings that count as the legal equivalent of a handwritten signature. The regulations for the use of electronic signatures are set out in 21 CFR Part 11 of the FDA. Each electronic signature must be assigned uniquely to one person and must not be used by any other person. It must be possible to confirm to the authorities that an electronic signature represents the legal equivalent of a handwritten signature. Electronic signatures can be biometrically based or the system can be set up without biometric features.

Encapsulation

The division of material into a hard gelatin capsule. Encapsulators should all have the following operating principles in common: rectification (orientation of the hard gelatin capsules), separation of capsule caps from bodies, dosing of fill material/formulation, rejoining of caps and bodies, and ejection of filled capsules

Endotoxin

A pyrogenic product (e.g., lipopolysaccharide) present in the bacterial cell wall. Endotoxin can lead to reactions in patients receiving injections ranging from severe fever to death.

Enzymes

Proteins that act as a catalyst in biochemical reactions.

Equipment Qualification

Qualification means the process to demonstrate the ability to fulfil specified requirements. EQ consists of proving and documenting that equipment or ancillary systems are properly installed (Installation Qualification, IQ), work correctly (Operations Qualification OQ), and the different sub-systems work together as a system (Performance Qualification PQ) and actually lead to the expected results.
Qualification is part of validation, but the individual qualification steps alone do not constitute a validated process.

Equipment Train

The sequence of equipment through which a product is produced or processed

Equipment Range

The full range that equipment is capable of performing, as per the manufacturer specification and tolerances. (a process may not utilize the full equipment range, operating over a narrower range)

ETO Sterilisation

Ethylene oxide is used for the sterilization of heat sensitive components and for sterilization of items that are difficult to sterilize using other methods (e.g. Autoclaving, Heat tunnels, Electron Beam irradiation etc.) Some materials may not be suitable or table when subjected to irradiation. Ethylene oxide sterilization is also used for the sterilization of medical devices in the final packs.

Excipient

Substances other than the API which have been appropriately evaluated for safety and are intentionally included in a drug delivery system to provide a specific role in manufacturing, shelf-life or physical property.

F

Factory Acceptance Testing (FAT)

An FAT or Factory Acceptance Test is an engineering activity that inspects and verifies that the equipment or system meets the requirements of the URS.

Failure Mode And Effects Analysis (FMEA)

A risk assessment tool that provides for an evaluation of potential failure modes and their likely effect on outcomes and/or product or process performance in order to prioritize risks and monitor the effectiveness of risk control activities. It is often used to identify areas within a given process, product, or system that render it vulnerable

FDA 483s

An FDA 483 letter typically includes a summary of findings and observations in relation to an audit or inspection where the FDA representatives have reason to believe GMP or other regulations have been violated or are not being met. In response to an FDA 483 letter, the company should address each item and provide a timeline for correction or request clarification of what changes are required

Fermentation

An anaerobic bioprocess. Fermentation is used in various industrial processes for the manufacture of products such as alcohols, acids, and cheese by the action of yeasts, molds, and bacteria. The fermentation process is used also in the production of monoclonal antibodies

Freeze Drying)

Lyophilization is the removal of ice or other frozen solvents from a material through the process of sublimation and the removal of bound water molecules through the process of desorption

Frank Pathogen

Microorganism responsible for infection in healthy individuals (i.e. individuals with normal operative and functional host defense mechanisms) that may bebacquired from exposure to other infected people or animals, environmental reservoirb(exogenous) or the individual's normal (endogenous) microbial flora. [PDA technical report 67]

Functional Design Specification (FDS)

A functional design specification is a document that specifies how particular requirements are met – this can be a combination of how the equipment/process operates mechanically/automatically etc. An FDS is typically written to response to a URS.

Fungicide

A disinfectant which kills fungi including their spores under defined conditions

Fluid

A fluid is a substance that undergoes continuous deformation when subjected to a shearing force.

G

GAMP (5)

Good Automated Manufacturing Practice (GAMP) is a set of guidelines for manufacturers and users of automated systems in regulated industries. Specifically, the Medical device, pharmaceutical and biopharmaceutical industries.
The application of GAMP and Validation of Automated Systems in manufacturing helps ensure that regulated medical devices and medicinal products have the required quality and are manufactured according to Good practices, meet regulatory and legal requirements and ensure patient safety.

Gene

The basic unit of heredity, which plays a part in the expression of a specific characteristic. The expression of a gene is the mechanism by which the genetic information that it contains is transcribed and translated to obtain a protein. A gene is a part of the DNA molecule that directs the synthesis of a specific polypeptide chain. It is composed of many codons. When the gene is considered as a unit of function in this way, the term cistron is often used.

Gene Transfer

The use of genetic or physical manipulation to introduce foreign genes into a host cells to achieve desired characteristics in progeny.

Gene therapy

An evolving technique used to treat genetic diseases. The medical procedure involves replacing, manipulating or supplementing non-functional genes with healthy genes so that they can function normally

Genetic disease

A disease or condition caused by a change or mutation in a gene, or a change in the chromosomes

Genetic engineering

The technique of removing, modifying or adding genes to a DNA molecule to change the information it contains. By changing this information, genetic engineering changes the type or amount of proteins an organism is capable of producing. Genetic engineering allows scientists to isolate a specific gene for a particular trait - such as resistance to insect attack - in a plant or animal, and transfer it into another plant

Genetic mapping

A research method that collects genetic information to determine the relative position of a gene or a phenotype in the genome

Genetic marker

A DNA sequence at a unique physical location in the genome, which varies sufficiently between individuals that its pattern of inheritance can be tracked through families and/or it can be used to distinguish among cell types. A marker may or may not be part of a gene. Markers are essential for use in linkage studies and genetic maps to help scientists to narrow down the possible location of new genes, and to discover the associations between genetic mutations and disease

Genetic modification

A general term which refers to any intentional change to the heritable traits of an organism. This includes both traditional breeding and recombinant DNA techniques

Genetic privacy

The freedom from unauthorized intrusion. Often referred to as the right to be let alone, it protects territorial, bodily, psychological and informational integrity and decision making. Many of these interests are directly implicated by genetic testing. Informational privacy protects the access, control and spread of personal information. Privacy is essential to maintaining relations of trust

Genetic testing

A laboratory test, done most often on a blood sample, but also on cheek cells, skin cells, bone marrow, amniotic fluid or a placenta sample. It looks at a particular gene for changes, or mutations, that might confirm the diagnosis of a genetic disease or that show a predisposition to a genetic disease

Genetic toxicology

A research field in which genetic samples from a living organism (including humans) are placed on a DNA microarray (gene chip) and tested in a computerized device for the presence of toxic substances from the environment. It is done to determine if the organism providing the sample has been exposed to specific chemicals which have caused problems such as mutations, cancer and birth defects. The study of the pattern of occurrence of such biomarkers in a sample of individuals or a community is called genetic epidemiology

Good Documentation Practices, GDP

The handling of written or pictorial information describing, defining, specifying and/or reporting of certifying activities, requirements, procedures or results in such a way as to ensure data integrit
Granulation

A process of creating granules. The powder morphology is modified through the use of either a liquid that causes particles to bind through capillary forces or dry compaction forces.

Grade A Areas

Aseptic processing areas, critical in nature where sterile products are exposed to the environment receiving no further sterilization. High-risk operations (for example aseptic stopperage, filling, loading of the lyophilizer) occur in Grade A areas. They are considered ISO 5 under both dynamic and static conditions.)

Grade B Areas

Aseptic processing areas where the sterile product is protected from the environment. Grade B processing areas are the background environments for Grade A areas and are considered ISO 7 environments in the dynamic state and ISO 5 environments under static conditions.

Grade C Areas

Non-critical areas where bulk product or materials are exposed to the environment, yet final sterilization has not yet been performed. Grade C areas are support areas for non-sterile production activities; purification, formulation, and preparation of components, equipment, etc. for sterilization. They are considered ISO 8 (Class 100,000) environments in the dynamic state and ISO 7 (Class 10,000) environments under static conditions.

Grade D Areas

Non-critical production areas, support areas, airlocks, or corridors. They are support areas for non-sterile production activities in closed systems; cell culture, or buffer and media preparation areas. Grade D Airlocks are used for the movement of product, materials, and personnel into classified areas.

GHTF

Global Harmonization Task Force

Glycoprotein

Protein to which groups of sugars become attached. Human blood group proteins, cell wall proteins and some hormones are examples of glycoproteins.

GxP

GxP is a general term for good practice with regard to quality guidelines and regulations. These guidelines are used in many fields, including the pharmaceutical, medical device and food industries. "x" is used as an umbrella letter representing different subjects or disciplines in industry. Some prime examples include GLP (Good Laboratory Practice), GDP (Good Documentation Practice), GEP (Good Engineering Practice) and GMP (Good Manufacturing Practices). Furthermore, the use of a lower case "c" as a prefix indicates "current" or "up-to-date"

H

Harm

Damage to health, including the damage that can occur from loss of product quality or availability.

High level risk assessment (HLRA)

A High level risk assessment that can be used at the beginning of a project to estimate the risk. Such as the risks involved with bringing in new computerised/automated equipment.

Hepatocyte

Any of the polygonal epithelial parenchymatous cells of the liver that secrete bile called also hepatic cell, liver cell

Heredity

The transfer of genetic information from parents to children

Heterozygote

An individual with two different alleles at a particular locus on a pair of chromosomes

Homozygote

An individual with two identical alleles at a particular locus on a pair of chromosomes

Hormones

A chemical that is made by one type of cell in the body and acts on another. Hormones act as messengers to tell the target cell to stop or start certain cellular processes

Host genomics

The genetic makeup of a person (host or patient)

HPLC

High Performance Liquid Chromatography (HPLC) - An instrumental separation technique used to characterize or to determine the purity of a BDP by passing the product (or its component peptides or amino acids) in liquid form over a chromatographic column containing a solid support matrix. The mode of separation, i.e. reversed phase, ion exchange, gel filtration, or hydrophobic interaction, is determined by the column matrix and the mobile phase. Detection is usually by UV absorbance or by electrochemical means.

HVAC

Heating, ventilation and air-conditioning (HVAC) systems are used to control the environmental conditions within an area or manufacturing facility. HVAC systems also provide comfortable conditions for operators based in the manufacturing environment. Temperature, relative humidity (RH) and ventilation should not adversely affect the quality of products during their manufacture and storage, or the proper functioning of equipment

Hydrogel

A biomaterial made up of a network of polymer chains that are highly absorbent and as flexible as natural tissue.

Hypothesis Testing

A structured approach that quantifies statistical confidence when making decisions based on data Different hypothesis tests are available, depending on the and type of data and goal of the testing.

I

ICH

International Conference on Harmonization of Technical Requirements for Registration of Pharmaceuticals for Human Use.

Intended Purpose
Intended purpose means the use for which the device is intended according to the data supplied by the manufacturer on the labelling, in the instructions and/or in promotional materials. (Chapter I section 1 of Annex IX of Directive 93/42/EEC)

Immunoassay

A qualitative or quantitative assay technique based on the measure of interaction of high affinity antibody with antigen used to identify and quantify proteins.

Immune system

A network of molecules, cells and organs that work together to protect the body against infection and disease

Immuno

Therapies and/or treatments that stimulate the immune system

Immunodeficiency

An innate, acquired, or induced inability to develop a normal immune response

Immunosuppression

The prevention or lessening of the immune response, for example, by irradiation or by administrating certain substances

Immunotoxicity

The toxicity of a therapeutic agent because it could cause immune reactions or allergy

Immunotyping

The process of screening patients specimens to identify the specific viral antigen on antigen presenting cells or detecting specific viral antibodies

Impurity

Any component of the new active pharmaceutical ingredient which is not the chemical entity defined as the new active pharmaceutical ingredient OR any component present in the active pharmaceutical ingredient or final product which is not the desired product, a product-related substance, or excipient including buffer components

Indirect Impact

Where a system is not expected to directly impact the product quality but supports or is ancillary to a direct impact system

Insulin

A hormone made by the pancreas that controls the level of sugar in the blood

Intellectual property

A form of creative endeavour that can be protected through a trademark, patent, copyright, industrial design or integrated topography. The patent system offers the only protection available for the intellectual products of research. There is increasing pressure on the patent system to patent life forms

Interferon

A protein first recognized in animals for its action in inhibiting viral replication and inducing resistance in host cells. The interferons (IFNs) are a highly conserved family of multi-functional, species-specific, secreted proteins originally classified on the basis of cellular origin including: leucocyte IFN (alpha), fibroblastic IFN (beta) and immune IFN (gamma). Examples from each of these classes have been cloned and commercialized.

The IFNs have been reclassified based on their recognition of cell-surface receptors. In humans there are three major classes: Type I consists mainly of the original types alpha (including various isoforms) and beta; Type II consists of IFN gamma; Type III consists of IFN lambda

The IFNs have multiple biological activities demonstrated to greater or lesser degree by the various types. These include: the induction of intracellular mechanisms having anti-viral effects (affecting viral protein synthesis) and anti-proliferative effects (affecting cell replication); the stimulation of cellular immune responses against viruses, bacteria and tumours; and, the regulation of immune and inflammatory responses

Invasive device

A device, which, in whole or in part, penetrates inside the body, either through a body orifice or through the surface of the body.

In Vitro

Biological reactions taking place outside the body in an artificial system.

In Vivo

Biological reaction taking place inside a living cell or organism.

Ion Exchange Chromatography (IEC)

A gradient driven separation based on the charge of the protein and its relative affinity for the chemical backbone of the column. Anion/cation exchange is commonly used for proteins.

IQ/OQ

Equipment IQ/OQ is defined as establishing documented evidence that all key aspects of the process equipment installation adhere to the manufacturer's approved specifications and any recommendations of the supplier of the equipment are suitably considered.
The process/equipment must also operate as intended and all user requirements are adequately fulfilled.

IFU

Instructions for Use.

(Plant) Injunction

An injunction is a judicial process initiated to stop or prevent violation of the law, such as to halt the flow of violative products in interstate commerce and to correct the conditions that caused the violation to occur. (FDA 21 U.S.C. 332; Rule 65, Rules of Civil Procedure).
If a firm has a history of violations and has promised correction in the past but has not made the corrections, the injunction is more likely to succeed. However, the freshness of the evidence is critical.

For an injunction action to be credible in the eyes of the Department of Justice (DOJ), the U.S. Attorney and the court, the evidence must be current. Timeliness is an important factor when considering an injunction action, with or without a Motion for Preliminary Injunction or a temporary restraining order (TRO). However, case quality and credibility must not be sacrificed to meet guideline time frames. The purpose of the guideline time frames is to limit, as much as can reasonably be expected, the need to update evidence. Updating entails extra work at all levels of the case development and review process and more importantly, delays obtaining an injunction which is intended to stop violations that adversely affect the safety or quality of products in commerce.

ISO

International Organization for Standardization. Agency responsible for developing international standards. E.g. ISO 13485 Medical Devices.

Isolator

A sealed enclosure, which provides full physical separation between the critical processing zone and the surrounding other processing zones. The internal surfaces of the isolator and of its contents are decontaminated, in accordance with defined objectives, by highly effective cycles. (e.g. Vaporised Hydrogen peroxide)

Enclosure capable of preventing ingress of contaminants by means of physical interior/exterior separation, and capable of being subject to reproducible interior bio-decontamination.

Isoelectric Precipitation

Isoelectric Precipitation works by reducing the electrostatic forces to near zero, allowing the proteins to precipitate out.

ISO 13485

ISO 13485, ISO standard, published in 2003, that represents the requirements for a comprehensive management system for the design and manufacture of medical devices.

ISO 14971

An ISO standard, published in 2007, that provides a framework and requirements for a risk management system for medical devices. This standard establishes the requirements for risk management to determine the safety of a medical device by the manufacturer during the product life cycle.

ISO 9001 - ISO 9001 is an ISO standard that represents the requirements for quality management systems. It is used across industries and is not specific to medical devices like ISO 13485

Item Master

A of all components that a manufacturer buys, builds or assembles into its products. The item master includes information like the size, shape, material, manufacturer, manufacturer part number and vendor for each component.

IVD

In Vitro Diagnostic tests are medical devices intended to perform diagnoses from assays in a test tube, or more generally in a controlled environment outside a living organism.

IVDD

The In Vitro Diagnostic Device Directive delineates requirements that in vitro diagnostic devices must meet before they can be sold in the EU market.

Intermediate

A material produced during steps of the processing of an API that undergoes further molecular change(s) or purification before it becomes an API.

J

JIT (Just in time)

A strategy used to monitor inventory levels with the goal of reducing inventory and associated carrying costs.

K

Kaizen

Kaizen is the Japanese word for improvement though the term is often meant to infer not only improvement but "continuous improvement". Although the event can be technical in nature, the key components of a Kaizen event involve copper-fastening employee involvement along with the right attitude and a culture that supports improvement and lean principles. In practical terms, Kaizen can be described as a highly focused "assault" on process to problem in order to realise a rapid improvement.

Kanban

A scheduling system that advises manufacturers what to produce, when to produce and how much to produce. Pioneered by Toyota, the approach is based on demand. Inventory is replenished only when visual cues like an empty bin, trolley or cart show that it's needed

Kaye (Validator)

A thermal measurement system used to record temperature and/or pressure of equipment. Kaye Validators are used in the equipment and process validation of autoclaves, freezers, fridges and sterilization tunnels (Depyrogenation tunnels)

L

Limulus Amoebocyte Lysate Test (LAL)

A sensitive test for the presence of endotoxins using the ability of the endotoxin to cause a coagulation reaction in the blood of a horseshoe crab. The LAL test is easier, quicker, less costly and much more sensitive that the rabbit test, but it can detect only endotoxins and not all types of pyrogens and must therefore be thoroughly validated before being used to replace the USP Rabbit Pyrogen test. Various forms of the LAL test include a gel clot test, a colormetric test, a chromogenic test, and a turbidimetric test.

Laminar flow

Laminar flow is when fluid particles move in parallel layers, at a constant velocity.

Lifecycle (Validation)

The Validation lifecycle refers to the requirement to control and document all validation activities from conception and URS stage to the retirement of equipment or a process. The lifecycle approach ensures compliance throughout the life of the process/equipment while maintaining a validated state throughout the application of change control.

Linearity

The ability of an analytical procedure (within a given range) to obtain test results that are directly proportional to the concentration (amount) of analyte in the sample.

Line Clearance

The act of performing and documenting the removal of materials from a production or packaging line and cleaning prior to the introduction of a new batch or lot.

Lyophilization (or Freeze Drying)

Lyophilization is the removal of ice or other frozen solvents from a material through the process of sublimation and the removal of bound water molecules through the process of desorption.

M

Mass Spectrometry

A technique useful in primary structure analysis by determining the molecular mass of peptides and small proteins. Often used with peptide mapping to identify variants in the peptide composition. Useful to locate disulfide bonds and to identify post- translational modifications.

Maximum Allowable Carry Over (MACO)

The amount of allowed product residue (carry-over) from lot-to-lot, batch-to-batch. This limit is based on the most conservative or lowest level of three MACO calculation methods (1) Limited based on Toxicity, (2) Limit based on Smallest Therapeutic Dose, and (3) Worst Case Dose.

Measurement Capability Index (MCI)

The Measurement Capability Index (MCI) represents the capability of the measurement system. It is used to evaluate the capability of the gauge to classify product against predetermined specifications.

Measurement System Analysis (MSA)

A study to determine the degree of error involved in measuring the given parameter. The measurement system involves the combination of operations, procedures, gauges, instruments, environmental conditions, people and software.

Medical Device

A medical device is "an instrument, apparatus, implement, machine, contrivance, implant, in vitro reagent, or other similar or related article, including a component part, or accessory which is:

•	recognized in the official National Formulary, or the United States Pharmacopoeia, or any supplement to them,
•	intended for use in the diagnosis of disease or other conditions, or in the cure, mitigation, treatment, or prevention of disease, in man or other animals, or
•	intended to affect the structure or any function of the body of man or other animals, and which does not achieve any of its primary intended purposes through chemical action within or on the body of man or other animals and which is not dependent upon being metabolized for the achievement of any of its primary intended purposes."

Medicinal Drug Products (Finished Products)

Finished dosage forms (e.g. tablet, capsule, or solution) that contain the active pharmaceutical ingredient usually combined with inactive ingredients. Medicinal products are intended to furnish pharmacological activity or other direct effect in the diagnosis, cure, mitigation, treatment, or prevention of disease or to affect the structure and function of the body.

MDD

The Medical Device Directive is intended to harmonize the laws relating to medical devices within the European Union. Medical Device Directive 93/42/EEC was most recently reviewed and amended by 2007/47/EC.

MHRA

The Medicines and Healthcare products regulatory Agency (MHRA) is the UK government agency which is responsible for ensuring that medicines and medical devices work and are acceptably safe.

Microorganism of Concern

A bacterium, yeast or mould that, due to it prominence in product recalls, infection outbreaks, nosocomial infections and the clinical literature, results in a multifactor risk assessment to determine whether the microorganism is objectionable or specified if it is present in a specific non-sterile product or atypical if it is present in a specific sterile product. [PDA TR 67]

Materials of Construction

Aka MOC, typically detailed certificates documenting the material properties and characteristics of materials used in equipment manufacture and fabrication. MOC certs are particularly import for product contact surfaces.

MSDS

Material Safety Data Sheet.

N

NCR

Non-Conformance Report.

NIH

National Institutes of Health (U.S.)

Noel

No Observed Effect Level. (in relation to Cleaning Validation)

No Impact

A system that does not directly impact product quality and does not support a direct impact system.

Non-conformity

A deficiency in a characteristic, product specification, CQA, process parameter, record, or procedure that renders the quality of a product unacceptable, indeterminate, or not according to specified requirements.

Non Parametric Data

Where the type of data is non variable Also referred to as attribute data eg (Visual inspection resulting in a PASS/FAIL result.

Notified Bodies

A notified body is a certification organisation which the national authority (the competent authority) of a member state designates to carry out one or more of the conformity assessment procedures or audits described in the annexes of the medical devices directives or GMP legislation.

NPI (New product introduction)

The market launch or commercialization of a new product. NPI takes place at the end of a successful product development project.

O

Open System

An environment in which system access is not controlled by persons who are responsible for the content of electronic records on the system (21 CFR, Part 11)

Opportunistic Pathogen

Microorganism responsible for infection in injured, invasively treated or immune-suppressed individuals that typically do not cause infection in healthy individuals, unlike frank pathogens. [PDA TR 67]

Outlier

A test result that is statistically different compared to a set of other test results obtained from the same sample or samples from the same lot of material.

Out-Of-Specification

A recorded result that falls outside the established specification(s) or acceptance criteria.

Out-Of-Trend

Analytical result, which is within specification or acceptance criteria, but different from those usually obtained or expected. Out-of-trend results should be investigated by the same general principles as out-of-specification results.

Quantitation limit

The lowest amount of analyte in a sample which can be quantitatively determined with suitable precision and accuracy for an analytical procedure. The quantitation limit is a parameter of quantitative assays for low levels of compounds in sample matrices and is used particularly for the determination of impurities and degradation products.

Overall Equipment Effectiveness(OEE)

A calculation for measuring the efficiency and effectiveness of a process, by Equipment breaking it down into three constituent components (the OEE Factors) Availability x Performance x Quality.

Overkill

Sterilization process that is demonstrated as delivering at least a 12 Spore Log Reduction (SLR) to a biological indicator having a resistance equal to or greater than the bioburden level.

P

Pan Coating

The uniform deposition of coating material onto the surface of a solid dosage form while being translated via a rotating vessel.

Parametric Release

A system of release that gives the assurance that the product is of the intended quality based on information collected during the manufacturing process and on the compliance with specific GMP requirements related to Parametric Release

Particle count test

Test covers verification of cleanliness. Dust particle counts measured. The number of readings and positions of tests should be defined in accordance with ISO 14644-1 Annex B5.

Passivation

Passivation can be described as the active chemical process used to obtain a uniform Chromium Oxide layer on Stainless Steel (SS) surfaces. The Chromium Oxide layer or film forms a protective coating that gives corrosion resistant properties.

PDCA (plan–do–check–act)

PDCA is a four-step management tool often used in GLP and GMP impacting environments. It introduces a repeatable and structured process-approach to solving problems and helps to drive consistent practices.

Plan
The plan step is used to establish the objectives and desired goals of the proposed changes or modifications. Documenting these goals is important as it will drive all aspects of the next steps in the PDCA process. Plan may also involve defining the problem or issue at hand and the specific tasks or resolutions required to rectify the issue.

Do
Step two-implement the plan and the changes identified in the initial step. The "do" step may require data collection and/or analysis prior to the implementation of changes. Training may also be required. Responsibilities should be clearly defined.

Check

Review results and analysis against the planned and expected results or goals. This "check" may simply not be confined to reviewing initial results, it may require monitoring over an extended period of time.

Act

The act step ensures if any further corrective actions or modifications that are noticed in the check step, the process will require the person to "act" on the findings. However, any proposed changes are better captured by returning to the first step and starting the process, either way, the application of PDCA will drive continuous improvement and issue the problem is fully addressed.

Pharmacokinetics

The study of how drugs are absorbed, distributed and cleared from the body

Phenotype

A set of observable physical characteristics of an organism

Plant genetics

The study of genetics in plants

Plant Molecular Farming (PMF)

This technique involves using genetically modified plants to produce substances that the plants typically do not produce naturally, such as industrial compounds or therapeutics

Plasmid

A DNA structure that is separate from the cell's genome and can replicate independently of the host cell. Plasmids are used in the laboratory to deliver specific DNA sequences into a cell

Plasticity

The ability of adult-derived stem cells to be capable of developing into cells types outside of the tissue of origin (for example, human blood stem cells have been shown to differentiate into liver cells

Polymerase Chain Reaction (PCR)

In vitro technique for amplifying nucleic acid. The technique involves a series of repeated cycles of high temperature denaturation, low temperature oligonucleotide primer annealing and intermediate temperature chain extension. Nucleic acid can be amplified a million- fold after 25- 30 cycles.

Peptide Mapping

A technique which involves the breakdown of proteins into peptides using highly specific enzymes. The enzymes cleave the proteins at predictable and reproducible amino acid sites and the resultant peptides are separated via HPLC or electrophoresis. A sample peptide map is compared to a map done on a reference sample as a confirmational step in the identity profiling of a product. It is also used for confirmation of disulfide bonds, location of carbohydrate attachment, sequence analysis, and for identification of impurities and protein degradation.

Performance indicators

Measurable values used to quantify quality objectives to reflect the performance of an organization, process or system, also known as performance metrics in some regions. (ICH Q10)

Performance Qualification (PQ)

Establishing by documented evidence that the process, under anticipated (controlled) conditions, consistently produces a product which meets predetermined requirements.

Poka Yoke

Poka Yoke is a technique for avoiding simple human error in the workplace. Simply put, it aims eliminate mistakes and is often referred to as mistake-proofing or fail-safe work methods. Poka Yoke is a system designed to prevent inadvertent errors made by workers performing a process. The word "Poka-Yoke" is Japanese for mistake-proofing or mistake avoidance. It involves the design of products, work practices, fixtures and jigs etc. that prevent the mistakes or errors that result in defects. A secondary aim of Poka Yoke is to make any defect easy to recognise with minimum time, skill and expertise. It is accepted as a simple and inexpensive way of preventing defects from being made or identifying a defect so that it is not passed to the next operation, downstream process and ultimately, the consumer.

Precision
The degree of agreement (scatter) between a series of measurements when a method is applied repeatedly to multiple samplings of a homogeneous sample or artificially prepared sample under the prescribed conditions. There are three types of precision; repeatability, intermediate precision and reproducibility.

Pressure cascade

A process whereby air flows from one area, which is maintained at a higher pressure, to another area at a lower pressure.

Protein

A polypeptide consisting of amino acids. In their biologically active states, proteins function as catalysts in metabolism and, to some extent, as structural elements of cells and tissues.

Pyrogenicity

The tendency for some bacterial cells or parts of cells to cause inflammatory reactions in the body, which may detract from their usefulness as pharmaceutical products.

Piping & Instrument Diagrams (P&IDs)

Engineering technical drawings that provide details of the connections and integration of equipment, services, material flows, plant controls and alarms. The P&ID also provide the reference for each tag or label used for identification.

PMA

Premarket approval by FDA is the required process of scientific review to guarantee safety and effectiveness for Class III devices.

PMDA

The Pharmaceutical and Medical Devices Agency in Japan reviews applications for marketing approval of pharmaceuticals and medical devices. It also monitors their post-marketing safety and provides relief compensation for people who have suffered from adverse drug reactions from pharmaceuticals or infections from biological products.

PMS

Post Marketing Surveillance is the practice of monitoring a pharmaceutical drug or device after it has been released on the market.

Process design

Defining the commercial manufacturing process based on knowledge gained through development and scale-up activities.

Process qualification

Confirming that the manufacturing process as designed is capable of reproducible commercial manufacturing.

Process window

The selected operating range of machine setting/parameter that will produce product to meet all quality and product specifications.

Product Recovery

Product recovery is a critical and important step in the process. It is also referred to as "Downstream processing". It is often the most expensive step in the process. For recombinant-DNA derived products, purification can often account for 90% of the total production costs.

Prospective Validation

Prospective Validation is when validation is done in advance of commercial manufacturing.

Procedures

Also known as Standard Operating Procedures, or SOPs, give directions for performing certain operations.

Protocols

Give instructions for performing and recording certain discreet operations. (Examples include engineering protocols, validation protocols etc.)

Pure

A term typically used within pharmaceutical manufacturing, a product or substance is pure if it is free of contaminants, foreign matter, chemicals and harmful microbes.

Q

QMS

Quality Management System can be expressed as the organizational structure, procedures, processes and resources needed to implement quality management.

Quality

The degree to which a set of inherent properties of a product, system, or process fulfils requirements. (ICH Q9)

Quality by design

This is a systematic approach that begins with predefined objectives and emphasizes product and process understanding and process control, based on sound Science and engineering principles.

Quality Management System

A Quality Management System, often abbreviated to (QMS) is any system based on a collection of business processes that are primarily focused on providing safe and quality products that consistently meet customer requirements.

Quarantine

The status of materials isolated physically or by other effective means pending a decision on their subsequent approval or rejection.

(Quality) Policy

A document in which a company or organization outlines their commitment and approach to quality. It usually sets out how they plan to achieve a high and consistent standard of quality. It should in some way speak to the customer or end user.

Qualification Plan

A Qualification Plan (QP) describes all the qualification measures and at which stage of the qualification the verification will be completed. It typically contains detailed descriptions of the necessary test measures and a description of the interdependencies of the individual tests. In some instances, there may not be a need or a requirement for a qualification plan. A validation plan can also serve to detail the qualification strategy.

QP

Companies that intend to manufacture or import medicinal products or intermediate products, for use in clinical trials or for market within the EU, must appoint the service of a Qualified Person, in order to comply with EU Good Manufacturing Practice Standards.

QPM

Quality Policy Manual.

QSP

Quality System Procedure.

QSR
Quality System Regulations.

R

Range

Range is defined as the interval between the upper and lower measurements required. The minimum specified range should be within the equipment range and validated to operate at all points within the range.

Recall

As defined at 21 CFR 7.3(g), "recall means a firm's removal or correction of a marketed product that the Food and Drug Administration considers to be in violation of the laws it administers and against which the agency would initiate legal action, 21 CFR 806.2(h). e.g., seizure. Recall does not include a market withdrawal or a stock recovery." Recall does not include routine servicing. Recall also does not include an enhancement, as defined by this guidance.

Relative humidity

The ratio of the actual water vapour pressure of the air to the saturated water vapour pressure of the air at the same temperature expressed as a percentage. More simply put, it is the ratio of the mass of moisture in the air, relative to the mass at 100% moisture saturation, at a given temperature.

Reusable medical device

A device intended for repeated use either on the same or different patients, with appropriate decontamination and other reprocessing prior to re-use.

Reusable Surgical Instrument

Instrument intended for surgical use by cutting, drilling, sawing, scratching, scraping, clamping, retracting, clipping or similar surgical procedures, without connection to any active medical device and which are intended by the manufacturer to be reused after appropriate procedures for cleaning and/or sterilisation have been carried out.

Re-Qualification

Requalification is designed to verify and ensure that the equipment/instrument/system is maintained in a qualified state after modification or after a stipulated time period (downtime).

Residue

Substance left on surfaces of equipment after cleaning that may pose as risk for subsequent use. Example: residues that may require cleaning include: product, excipients, raw materials/intermediates, non-volatile solvent, non-intrinsic cleaning agents such as detergents, etc.

Residual Risk

The risk level remaining after applying the identified controls on a high risk of harms and hazards manifestation.

Resolution

The smallest change in quantity that can be detected or provided by an instrument.

Residual Solvent

Organic volatile chemicals used or produced during the manufacture of APIs or excipients, or in the preparation of medicinal products.

Retain Samples

Samples that are kept for potential investigations and retests. It should be noted that retain samples are not a regulatory requirement, per Annex 10 or 21 CFR part 11.

Retrospective Validation

Retrospective validation is used for facilities or processes that have not completed formal validation. Historical data or a retrospective review can provide the evidence that the process or facility is operated as intended.

Rinse Sampling

Using a solvent to contact all surfaces of the sampled item to quantitatively remove target residue. The solvent can be water, water with pH adjusted, or organic solvent.

Right First Time

Right First Time strives to create a culture of excellence. People are challenged with performing their tasks always in the correct manner to achieve the correct results always - right the first time.

Risk

The combination of the probability of occurrence of harm and the severity of that harm.

Risk Analysis

The use of available information to identify hazards and estimate risk.

Risk Management

Risk management involves the systematic application of management policies, practices and procedures that identify, analyse, control and monitor risk.

It is important to recognise that risk management should begin at the outset of the design and development phase of a project. The first step is to identify the user needs and intended use and application of the device.

RoHS

"Restriction of Hazardous Substances in electrical and electronic equipment 2002/95/EC". An initiative that was adopted by the European Union (EU) in February 2003 and put into effect July 1, 2006
Ruggedness

An indication of how resistant a test method or process is to typical variations in operation, such as those to be expected when using different analysts, different instruments and different reagent batches.

Rouging

Rouging is a form of corrosion found in stainless steels. It can be due to iron contamination of the stainless steel surface due to welding of non-stainless steel for support columns, or other temporary means, which when welded off leaves a low chromium area

S

Scaffold

A structure of artificial or natural materials on which tissue is grown to mimic a biological process outside the body.

SDA (Plates)

Sabouraud agar is a type of agar growth media containing peptones. It is used to cultivate dermatophytes and other types of fungi

Skid

A modular process that can be plugged into a process onsite, with little construction or integration. Skids are used as part of Clean-In-Place solutions within Food and Beverage, and Pharmaceutical industries.

SKU

(Stock keeping unit) A unique sales stock identifier.

Southern Blot

Technique for transferring DNA fragments from an agarose gel to a nitrocellulose filter on which they can be hybridized to a complementary DNA

Specifications

A approved document detailing the requirements with which the products or materials used or obtained during manufacture have to conform. They serve as a basis for quality evaluation.

Specified Microorganism

Microorganism with limit tests for absence in 1 or 10 g/mL of a finished product, as described in USP<62>/EP 2.6.23 and USP<1111>/EP 5.1.4. [PDA TR 67]

Specificity

The ability to assess unequivocally the analyte in the presence of components, which may be expected to be present.

Spore Log Reduction

The actual logarithmic reduction of the Biological Indicator (BI) population achieved during the cycle. Sterilisation The actual logarithmic reduction of the BI population achieved during the cycle.

Stability

Stability studies are used to demonstrate and justify assigned expiration or retest dates.

5S

5S is a Japanese methodology of organising and storing items in a work or lab environment. It has been adopted by many Western companies as a tool to help maintain standards and reduce errors and mix-ups. The "5s" represents each stage of the method.

Standard work

Standard work or standardised work is a particular type of work instruction. It is also a "lean" (see definitions and acronyms) tool as it not only creates a baseline but aims to create a balanced work flow with optimum product output.

Software Requirement Specification, SRS

An SRS can be written to interpret the requirements of a URS and how they relate to the requirement or how the requirement is met in practical terms regarding software

Sort

Sorting out any items that are not in use and removing to a more appropriate area or to storage or the bin.

Set-in-Order

The idea of "Set-in-Order" is to be always organised. "A place for everything and everything in its place. "If we "set-in-order" we can help to make live processing and testing more efficient and reduce the risk of errors, omissions and accidents.

Shine

Regular cleaning is an important practice and it is always helpful to "Clean as you go."

Standardise

Implement standard practices through SOPs and training. Standardisation can also be applied to work station layout.

Sustain

Make it a habit! After implementing a 5s methodology, it is only effective if continuous efforts are made to "sustain" the changes.

Sequencing of DNA Molecules

The process of finding the order of nucleotides (guanine, adenine, cytosine and thymine) that make up a DNA or RNA fragment

Sex chromosome

The 23rd pair of chromosomes in humans are the sex chromosomes. Females have two X chromosomes and males have an X and a Y chromosome

Single nucleotide polymorphism (SNP)

Individual differences at a single nucleotide of DNA. This genotypic difference can cause a phenotypic difference in hair colour, height or response to a drug, depending on the gene

Somatic cell

Any cell in the body except the germ cells (egg and sperm)

Somatic cell nuclear transfer (SCNT)

A cloning technique where the nucleus from an unfertilized egg is removed and replaced with the nucleus from a somatic cell. The resulting egg will carry the full complement of genetic material of the host organism. This is how Dolly the cloned sheep was produced; she was genetically identical to her "mother". This technique can be used both for reproductive cloning and therapeutic cloning

Stem cell

A fundamental cell that has the potential to develop into any of the 210 different cell types found in the human body. Human life begins with stem cells, which divide again and again and branch off into special roles, like becoming liver or heart cells. They are an important resource for disease research and for the development of new ways to treat disease

Stem cell differentiation

The process by which a stem cell can become a specific cell type. Stem cell differentiation begins when they are exposed to certain biochemical cues - whether physiological or experimental. Biochemical cues in different parts of the body stimulate stem cells to grow into the specific cells needed in that location

Sterility Assurance (SAL)

Probability of a single viable microorganism occurring on an item after sterilization. For a terminally sterilized medical device to be designated as "sterile", the minimum sterility assurance level shall be SAL = 10^{-6} or better. When applying this quantitative value to assurance of sterility, an SAL of 10^{-7} has a lower value but provides a greater assurance of sterility than an SAL of 10^{-6}

T

Tableting

The reconstitution of a powder blend in which compression force is applied to form a single unit dose. (tablet)

Tableting press

Tablet press subclasses primarily are distinguished from one another by the method that the powder blend is delivered to the die cavity. Tablet presses can deliver powders without mechanical assistance (gravity), with mechanical assistance (automation), by rotational forces (centrifugal), and in two different locations where a tablet core is formed and subsequently an outer layer of coating material is applied (compression coating).

Therapeutic Dose

Acceptable quantity of a medicinal product to be administered expressed in mg per patient.

Total organic carbon (TOC)

Total organic carbon (TOC) analysis is a fast and effective analytical test method used for cleaning verification and validation in pharmaceutical manufacturing. It is used to test for residues of previously manufactured products (actives and excipients), cleaning detergents, chemicals, solvents, degradants and microbial contaminants

Traceability Matrix

A Traceability Matrix is a document that links the user requirements and specifications to where the verification and testing has been documented within the validation activities.
A traceability matrix illustrates that all user requirements are traceable to the evidence based test.

Trait

A characteristic of an organism

Transcription

A process in the cell where the DNA is used as a template to make the messenger RNA
Transfer RNA (tRNA)

RNA molecules that bind to amino acids and carry them to the ribosomes where proteins are made

Transformation

A process by which the genetic information of an organism is changed by the addition of foreign DNA

Transgenics

The insertion or splicing of specific genetic sequences from one species into the functioning genome of an unrelated species to transfer desired properties for human purposes. This may be viewed as a more precise form of hybridization or plant/animal breeding, with the added consideration that genetic material from species significantly different from one another is involved (for example, the insertion of genetic material from an animal into a plant or vice versa). Another possibility is the transfer of genetically controlled properties between different animal species, such as the breeding of goats whose milk yields spider silk for possible development of new structural materials

TSA (Plates)

Trypticase soy agar or tryptone soya agar (TSA) are growth media for the culturing of bacteria. They are general-purpose, nonselective media providing enough nutrients to allow for a wide variety of microorganisms to grow

Turbulent flow

Turbulent flow is when the movement of fluid particles are varying in velocity and direction.

T- Helper Cells

T- lymphocytes with the specific capacity to help other cells, such as B- lymphocytes, to make antibodies. T-helper cells are also required for the induction of other T- lymphocyte activities. Synonym is T inducer cell, T4 cell, or CD 4 lymphocyte.

T- Suppressor Cells

T- lymphocytes with specific capacity to inhibit T-helper cell function.

Transcription

The first stage in the expression of a gene by means of genetic information being transmitted from the DNA in the chromosomes to messenger RNA.

Translation

The second stage in the expression of a gene by means of genetic information being transmitted from the mRNA to the synthesis of protein.

U

Unidirectional Air Flow

Air flow moving in a single direction, in a robust and uniform manner, and at sufficient speed to reproducibly sweep particles away from the critical processing or testing area (FDA)

Uniform

The product is manufactured consistently and will have the same quality between batches manufactured on different days.

UDI, Unique Device Identification

The UDI is a series of numeric or alphanumeric characters that is created through a globally accepted device identification and coding standard. It allows the unambiguous identification of a specific medical device on the market.

Uninterrupted Power Supply

An uninterruptible power supply (UPS) is a system for buffering the main power supply. If the power supply fails, the battery of the UPS supplies the required power. When the power supply returns, the UPS battery stops supplying power and is recharged.

Unit Operation

Unit operations are the individual steps in the process that modify materials and their properties at each step of the process. Each unit operation comes together to create a complete process.

User Requirement Specification, URS

The URS is a critical document that defines the requirements of a particular system, equipment or process. Requirements such as the functional and operational aspects of the system are typically documented here.

USP

United States Pharmacopoeia.

UV Spectroscopy

A quantitation technique for proteins using their distinctive absorption spectra due to the presence of side- chain chromophores (phenylalanine, tryptophan, and tyrosine). Since this absorbance is linear, highly purified proteins can be quantitated by calculations using their molar extinction coefficient.

V

Validation

Validation is confirmation via documented evidence that the particular requirements for a specific intended use can be consistently fulfilled under anticipated conditions.

Validation Master Plan

A document providing information on a company's validation work programme. It typically details timescales for the validation work to be performed along with the key deliverables.

Verification

Verification confirmation by examination and provision of objective evidence (i.e. documentation) that the specified requirements have been fulfilled.

Vaporized Hydrogen Peroxide (VHP)

Vaporization of liquid hydrogen peroxide which results in a mixture of VHP and water vapor. The VHP mixture is used to decontaminate isolators.

Visibly Clean

Free of any residue that is visible to a person with normal eyesight under adequate lighting conditions.

Vaccine

A preparation that contains an agent or its components, administered to stimulate an immune response that will protect a person from illness due to that agent. A therapeutic (treatment) vaccine is given after disease has started and is intended to reduce or arrest the progress of the disease. A preventive (prophylactic) vaccine is intended to prevent disease from starting. Agents used in vaccines may be whole-killed (inactive), live-attenuated (weakened) or artificially manufactured. It can be created using the recombinant DNA process

Vector

A vehicle that carries foreign genes into an organism and inserts them into the organism's genome. Modified viruses are used as vectors for gene therapy

Virus

A submicroscopic particle that can infect other organisms. It cannot reproduce on its own but infects an organism's cell in order to use that cell's reproductive machinery to create more viruses. It usually consists of a DNA or RNA genome enclosed in a protective protein coat

W

Water Activity (Aw)

Ration of the vapor pressure of water in a product to the vapor pressure of pure water at the same temperature; numerically equal to 1/100 of the relative humidity (RH) generated by the product in a closed system and is a measure of the free or available moisture in the material. (PDA TR 67)

Warning Letter

A warning letter is a correspondence that notifies regulated industry about violations that FDA has documented during its inspections or investigations.

WEEE Directive

Waste electrical and electronic equipment directive. European Community directive 2002/96/EC where manufacturers are responsible for disposing of electrical/electronic waste.

Western Blot

This test is used to detect contaminating cell substrates and to evaluate recombinant polypeptides. After electrophoretic separation, the negatively charged proteins (the antigens) are electrophoretically transferred from the polyacrylamide gel onto a nitrocellulose membrane positioned on the anode side of the gel. Following incubation of the membrane with a specific antibody, they are labelled with another anti-antibody for detection.

WFI (Water for injection)

WFI is sterile and pyrogen free water containing o less than 10 CFU/100ml (Colony Forming Units) with a sample size of between 100 and 300 ml and an endotoxin level < 0.25 EU/ml.

WHO

World Health Organization.

WI

Work Instructions.

Witnessed By

When signed or initialed is legal proof that the individual signing is physically present and observes the step, calculation, or operation being performed by someone else, and that all entries of data are true and accurate.

Worst Case

A set of conditions or parameters which, in combination with product specification or attributes at their limits, pose the greatest challenge to the process.

X

--

Y

--

Z
Zone Classification

Zone classification refers to GMP areas which include controlled (aka classified) and non-controlled manufacturing areas. Areas may be classified based on EU Grades A–D and/or ISO Class 5–8 (in the US - Class 100–Class 100,000 areas.

www.ingramcontent.com/pod-product-compliance
Lightning Source LLC
Chambersburg PA
CBHW061441180526
45170CB00004B/1510